All About Me

Places I Go

By Nick Rebman

level 1
little blue readers

www.littlebluehousebooks.com

Little Blue House is distributed by North Star Editions:
sales@northstareditions.com | 888-417-0195

Produced for Little Blue House by Red Line Editorial.

Photographs ©: Monkey Business Images/Shutterstock Images, cover, 12–13; weedezign/Shutterstock Images, 4; wavebreakmedia/Shutterstock Images, 7; imtmphoto/Shutterstock Images, 8–9; Kzenon/Shutterstock Images, 11; LightField Studios/Shutterstock Images, 15; Kapi Ng/Shutterstock Images, 16 (top left); canbedone/Shutterstock Images, 16 (top right); Zynatis/Shutterstock Images, 16 (bottom left); Sorbis/Shutterstock Images, 16 (bottom right)

Library of Congress Control Number: 2019908245

ISBN
978-1-64619-003-4 (hardcover)
978-1-64619-042-3 (paperback)
978-1-64619-081-2 (ebook pdf)
978-1-64619-120-8 (hosted ebook)

Printed in the United States of America
Mankato, MN
012020

About the Author

Nick Rebman enjoys reading, walking his dog, and traveling to places where he doesn't speak the language. He lives in Minnesota.

Table of Contents

Places I Go

I go to many places.

I go to school.

I go to many places.

I go to the library.

I go to many places.

I go to the store.

I go to many places.

I go to the park.

I go to many places.

I go to the pool.

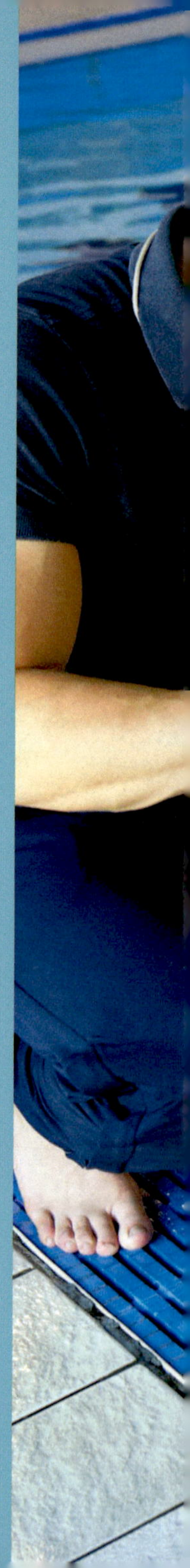

I go to many places.

I go to bed.

Glossary

library

pool

park

store

Index